LET US ADD THOSE NUMBERS

Name- _____ Class- _____ Date- _____

1
$$
\begin{array}{r}
1 \\
+3 \\
\hline
\end{array}
$$

2
$$
\begin{array}{r}
4 \\
+2 \\
\hline
\end{array}
$$

3
$$
\begin{array}{r}
5 \\
+3 \\
\hline
\end{array}
$$

4
$$
\begin{array}{r}
4 \\
+1 \\
\hline
\end{array}
$$

5
$$
\begin{array}{r}
4 \\
+2 \\
\hline
\end{array}
$$

6
$$
\begin{array}{r}
4 \\
+3 \\
\hline
\end{array}
$$

7
$$
\begin{array}{r}
5 \\
+4 \\
\hline
\end{array}
$$

8
$$
\begin{array}{r}
1 \\
+4 \\
\hline
\end{array}
$$

9
$$
\begin{array}{r}
4 \\
+4 \\
\hline
\end{array}
$$

LET US ADD THOSE NUMBERS

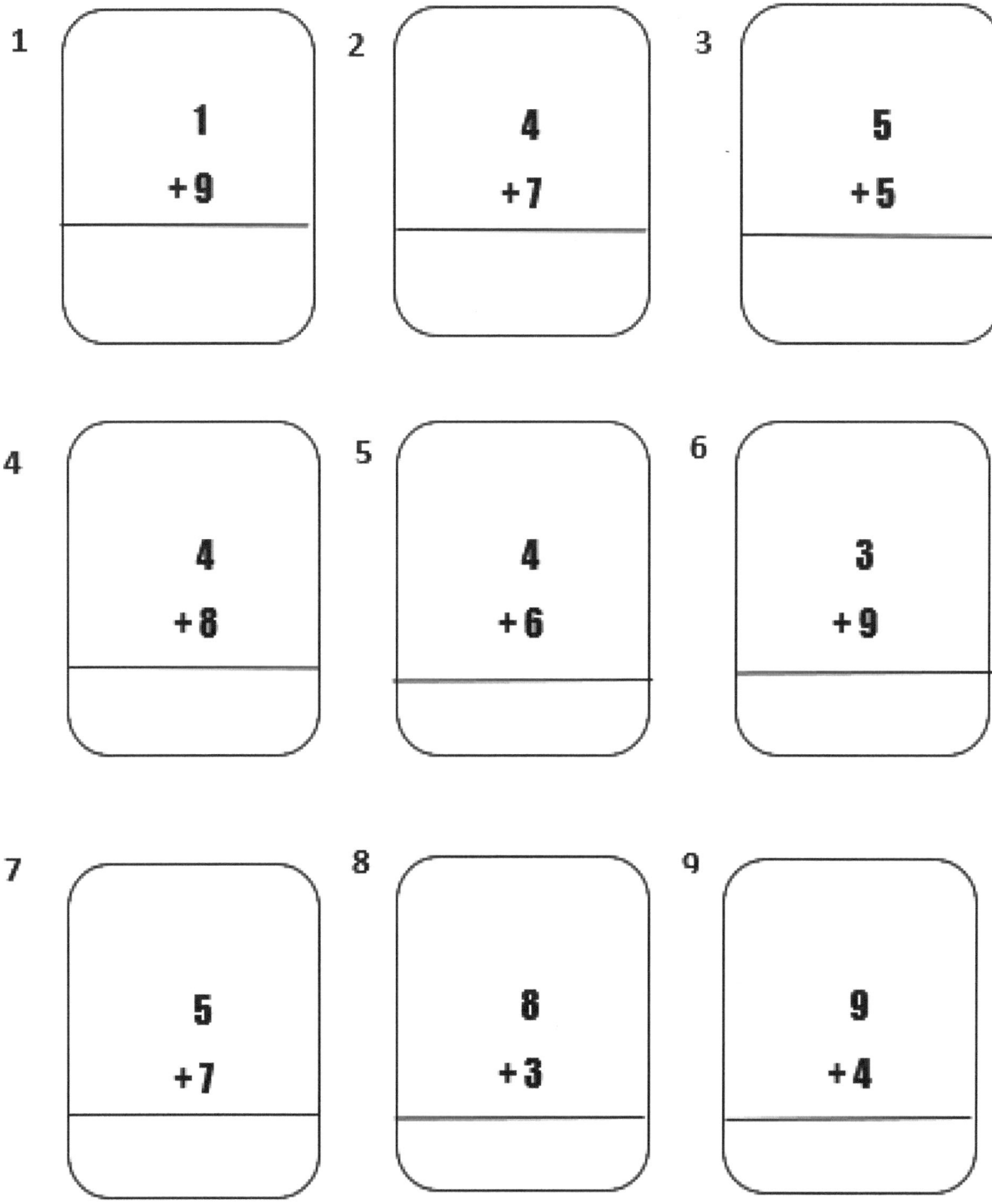

1.
```
  1
+ 9
```

2.
```
  4
+ 7
```

3.
```
  5
+ 5
```

4.
```
  4
+ 8
```

5.
```
  4
+ 6
```

6.
```
  3
+ 9
```

7.
```
  5
+ 7
```

8.
```
  8
+ 3
```

9.
```
  9
+ 4
```

LET US ADD THOSE NUMBERS

Name – _____ Date- _____ Score- _____

1
$$\begin{array}{r} 2 \\ +\,3 \\ \hline \end{array}$$

2
$$\begin{array}{r} 1 \\ +\,2 \\ \hline \end{array}$$

3
$$\begin{array}{r} 6 \\ +\,3 \\ \hline \end{array}$$

4
$$\begin{array}{r} 4 \\ +\,0 \\ \hline \end{array}$$

5
$$\begin{array}{r} 0 \\ +\,2 \\ \hline \end{array}$$

6
$$\begin{array}{r} 1 \\ +\,3 \\ \hline \end{array}$$

7
$$\begin{array}{r} 3 \\ +\,4 \\ \hline \end{array}$$

8
$$\begin{array}{r} 1 \\ +\,6 \\ \hline \end{array}$$

9
$$\begin{array}{r} 7 \\ +\,2 \\ \hline \end{array}$$

LET US ADD THOSE NUMBERS

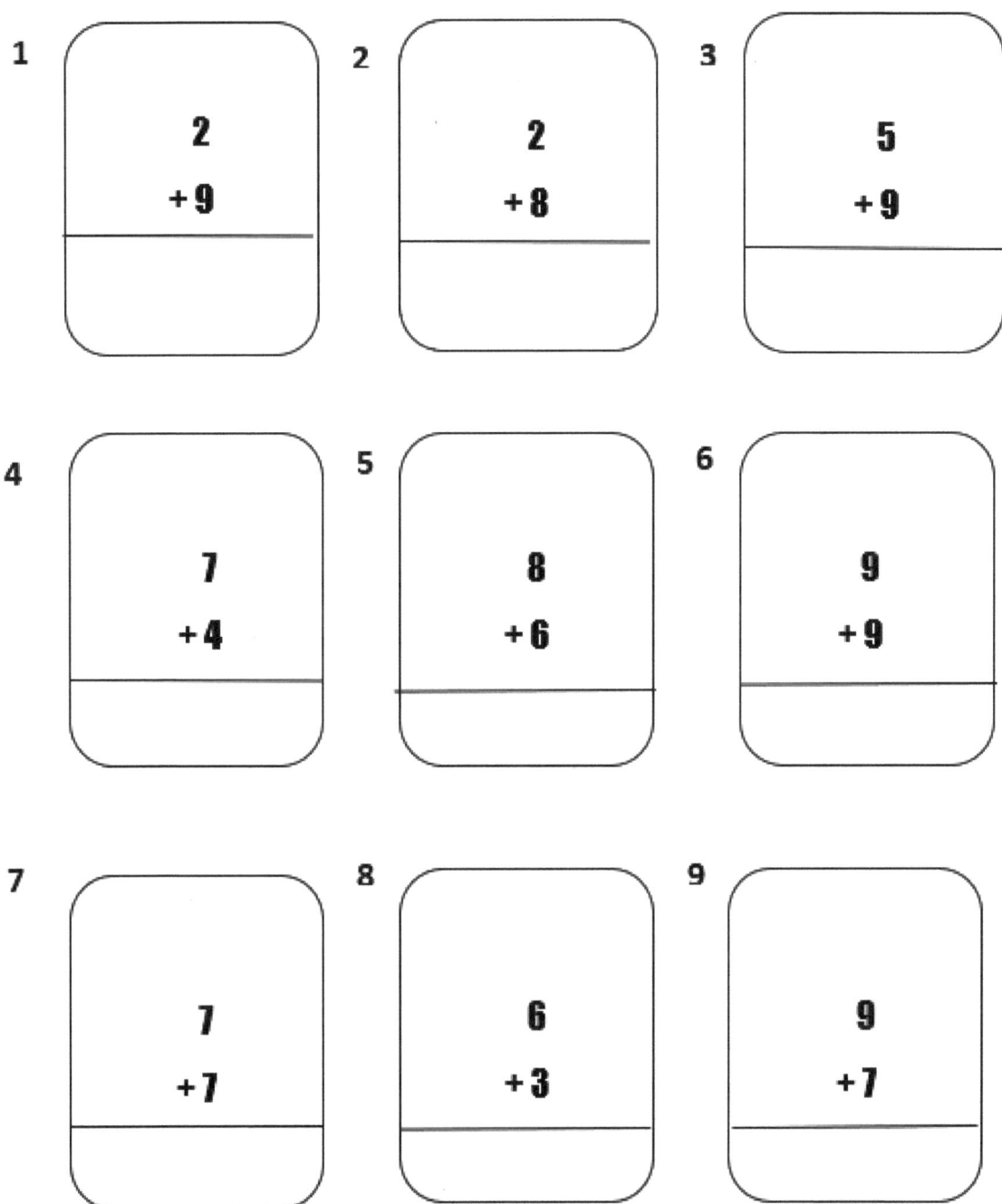

1
2
+9

2
2
+8

3
5
+9

4
7
+4

5
8
+6

6
9
+9

7
7
+7

8
6
+3

9
9
+7

LET US ADD THOSE NUMBERS

Name- _____ Class- _____ Date- _____

1
```
  1
 +8
```

2
```
  6
 +4
```

3
```
  3
 +2
```

4
```
  5
 +7
```

5
```
  9
 +2
```

6
```
  1
 +6
```

7
```
  8
 +3
```

8
```
  4
 +6
```

9
```
  2
 +9
```

LET US ADD THOSE NUMBERS

Name- _____ Class- _____ Date- _____

1
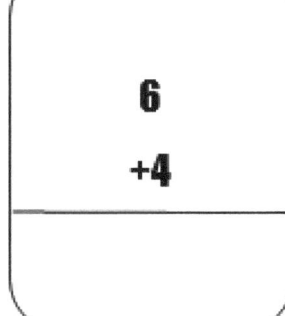
```
  1
 +8
____
```

2
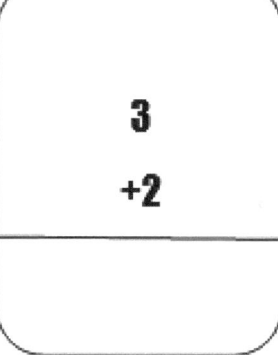
```
  6
 +4
____
```

3
```
  3
 +2
____
```

4
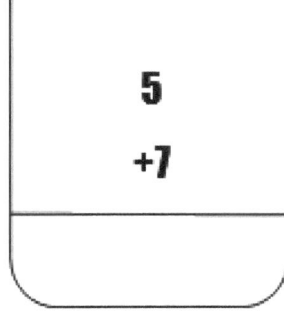
```
  5
 +7
____
```

5
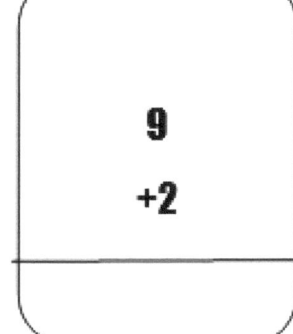
```
  9
 +2
____
```

6
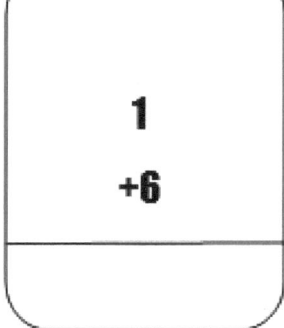
```
  1
 +6
____
```

7
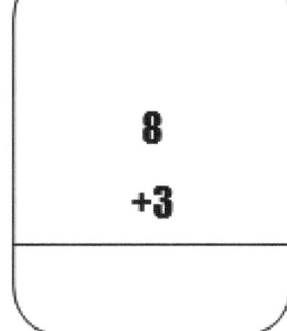
```
  8
 +3
____
```

8
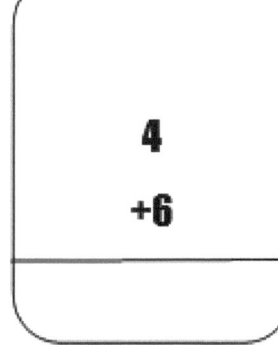
```
  4
 +6
____
```

9
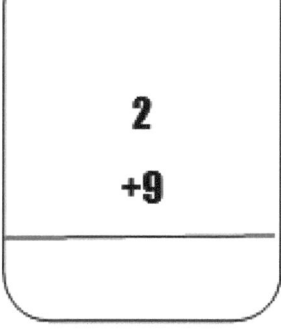
```
  2
 +9
____
```

LET US ADD THOSE NUMBERS

Name- _____ Class- _____ Date- _____

1.
$$\begin{array}{r} 5 \\ +1 \\ \hline \end{array}$$

2.
$$\begin{array}{r} 3 \\ +7 \\ \hline \end{array}$$

3.
$$\begin{array}{r} 6 \\ +1 \\ \hline \end{array}$$

4.
$$\begin{array}{r} 9 \\ +4 \\ \hline \end{array}$$

5.
$$\begin{array}{r} 8 \\ +6 \\ \hline \end{array}$$

6.
$$\begin{array}{r} 2 \\ +3 \\ \hline \end{array}$$

7.
$$\begin{array}{r} 1 \\ +4 \\ \hline \end{array}$$

8.
$$\begin{array}{r} 7 \\ +6 \\ \hline \end{array}$$

9.
$$\begin{array}{r} 5 \\ +9 \\ \hline \end{array}$$

LET US ADD THOSE NUMBERS

Name- _____ Class- _____ Date- _____

1
$$\begin{array}{r} 4 \\ +5 \\ \hline \end{array}$$

2
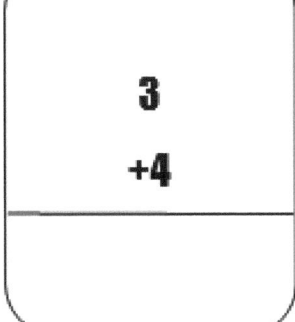
$$\begin{array}{r} 3 \\ +4 \\ \hline \end{array}$$

3
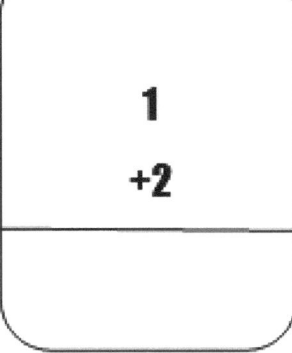
$$\begin{array}{r} 1 \\ +2 \\ \hline \end{array}$$

4
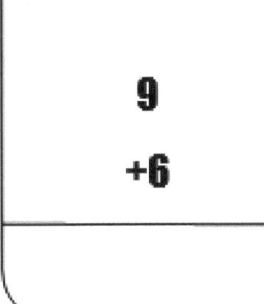
$$\begin{array}{r} 9 \\ +6 \\ \hline \end{array}$$

5
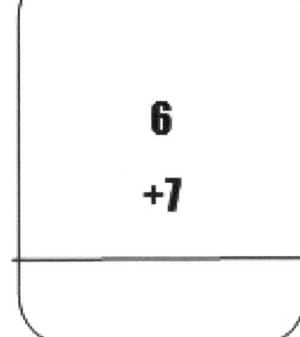
$$\begin{array}{r} 6 \\ +7 \\ \hline \end{array}$$

6
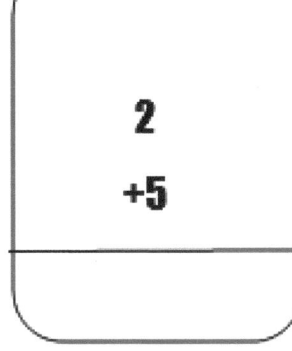
$$\begin{array}{r} 2 \\ +5 \\ \hline \end{array}$$

7
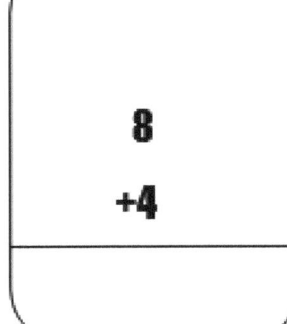
$$\begin{array}{r} 8 \\ +4 \\ \hline \end{array}$$

8
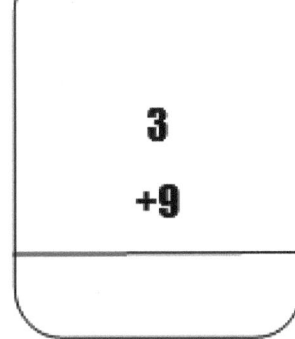
$$\begin{array}{r} 3 \\ +9 \\ \hline \end{array}$$

9
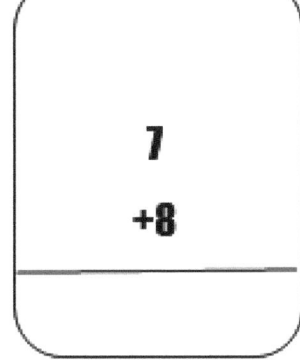
$$\begin{array}{r} 7 \\ +8 \\ \hline \end{array}$$

LET US ADD THOSE NUMBERS

Name- _____ Class- _____ Date- _____

1.
$$\begin{array}{r} 4 \\ +8 \\ \hline \end{array}$$

2.
$$\begin{array}{r} 3 \\ +9 \\ \hline \end{array}$$

3.
$$\begin{array}{r} 7 \\ +8 \\ \hline \end{array}$$

4.
$$\begin{array}{r} 1 \\ +3 \\ \hline \end{array}$$

5.
$$\begin{array}{r} 5 \\ +6 \\ \hline \end{array}$$

6.
$$\begin{array}{r} 2 \\ +6 \\ \hline \end{array}$$

7.
$$\begin{array}{r} 9 \\ +1 \\ \hline \end{array}$$

8.
$$\begin{array}{r} 4 \\ +3 \\ \hline \end{array}$$

9.
$$\begin{array}{r} 2 \\ +4 \\ \hline \end{array}$$

LET US SUBTRACT THOSE NUMBERS

Name- _____ Class- _____ Date- _____

1
$$\begin{array}{r} 6 \\ -2 \\ \hline \end{array}$$

2
$$\begin{array}{r} 9 \\ -3 \\ \hline \end{array}$$

3
$$\begin{array}{r} 8 \\ -1 \\ \hline \end{array}$$

4
$$\begin{array}{r} 7 \\ -5 \\ \hline \end{array}$$

5
$$\begin{array}{r} 9 \\ -2 \\ \hline \end{array}$$

6
$$\begin{array}{r} 5 \\ -3 \\ \hline \end{array}$$

7
$$\begin{array}{r} 6 \\ -4 \\ \hline \end{array}$$

8
$$\begin{array}{r} 7 \\ -1 \\ \hline \end{array}$$

9
$$\begin{array}{r} 9 \\ -6 \\ \hline \end{array}$$

LET US SUBTRACT THOSE NUMBERS

Name- _____ Class- _____ Date- _____

1

$$\begin{array}{r} 9 \\ -6 \\ \hline \end{array}$$

2

$$\begin{array}{r} 8 \\ -4 \\ \hline \end{array}$$

3

$$\begin{array}{r} 7 \\ -3 \\ \hline \end{array}$$

4

$$\begin{array}{r} 5 \\ -1 \\ \hline \end{array}$$

5

$$\begin{array}{r} 9 \\ -7 \\ \hline \end{array}$$

6

$$\begin{array}{r} 8 \\ -2 \\ \hline \end{array}$$

7

$$\begin{array}{r} 6 \\ -1 \\ \hline \end{array}$$

8

$$\begin{array}{r} 5 \\ -2 \\ \hline \end{array}$$

9

$$\begin{array}{r} 9 \\ -4 \\ \hline \end{array}$$

LET US SUBTRACT THOSE NUMBERS

Name- _____ Class- _____ Date- _____

1
$$\begin{array}{r} 6 \\ -3 \\ \hline \end{array}$$

2
$$\begin{array}{r} 8 \\ -6 \\ \hline \end{array}$$

3
$$\begin{array}{r} 9 \\ -1 \\ \hline \end{array}$$

4
$$\begin{array}{r} 5 \\ -4 \\ \hline \end{array}$$

5
$$\begin{array}{r} 7 \\ -6 \\ \hline \end{array}$$

6
$$\begin{array}{r} 8 \\ -3 \\ \hline \end{array}$$

7
$$\begin{array}{r} 9 \\ -5 \\ \hline \end{array}$$

8
$$\begin{array}{r} 6 \\ -2 \\ \hline \end{array}$$

9
$$\begin{array}{r} 7 \\ -4 \\ \hline \end{array}$$

LET US SUBTRACT THOSE NUMBERS

Name- _____ Class- _____ Date- _____

1

$$5$$
$$-1$$

2

$$9$$
$$-8$$

3

$$8$$
$$-5$$

4

$$6$$
$$-3$$

5

$$7$$
$$-2$$

6

$$9$$
$$-1$$

7

$$5$$
$$-3$$

8

$$8$$
$$-7$$

9

$$6$$
$$-5$$

LET US SUBTRACT THOSE NUMBERS

Name- _____ Class- _____ Date- _____

1.
$$\begin{array}{r} 9 \\ -2 \\ \hline \end{array}$$

2.
$$\begin{array}{r} 7 \\ -3 \\ \hline \end{array}$$

3.
$$\begin{array}{r} 5 \\ -1 \\ \hline \end{array}$$

4.
$$\begin{array}{r} 8 \\ -6 \\ \hline \end{array}$$

5.
$$\begin{array}{r} 6 \\ -2 \\ \hline \end{array}$$

6.
$$\begin{array}{r} 9 \\ -4 \\ \hline \end{array}$$

7.
$$\begin{array}{r} 7 \\ -5 \\ \hline \end{array}$$

8.
$$\begin{array}{r} 8 \\ -2 \\ \hline \end{array}$$

9.
$$\begin{array}{r} 6 \\ -1 \\ \hline \end{array}$$

LET US SUBTRACT THOSE NUMBERS

Name- _____ Class- _____ Date- _____

1
$$\begin{array}{r} 4 \\ -\ 2 \\ \hline \end{array}$$

2
$$\begin{array}{r} 3 \\ -\ 2 \\ \hline \end{array}$$

3
$$\begin{array}{r} 1 \\ -\ 1 \\ \hline \end{array}$$

4
$$\begin{array}{r} 4 \\ -\ 1 \\ \hline \end{array}$$

5
$$\begin{array}{r} 4 \\ -\ 3 \\ \hline \end{array}$$

6
$$\begin{array}{r} 4 \\ -\ 4 \\ \hline \end{array}$$

7
$$\begin{array}{r} 5 \\ -\ 4 \\ \hline \end{array}$$

8
$$\begin{array}{r} 1 \\ -\ 1 \\ \hline \end{array}$$

9
$$\begin{array}{r} 7 \\ -\ 4 \\ \hline \end{array}$$

LET US SUBTRACT THOSE NUMBERS

Name- _____ Class- _____ Date- _____

1.
$$5 - 2$$

2.
$$4 - 2$$

3.
$$4 - 1$$

4.
$$2 - 1$$

5.
$$5 - 3$$

6.
$$9 - 4$$

7.
$$7 - 1$$

8.
$$8 - 1$$

9.
$$9 - 3$$

LET US SUBTRACT THOSE NUMBERS

Name- _____ Class- _____ Date- _____

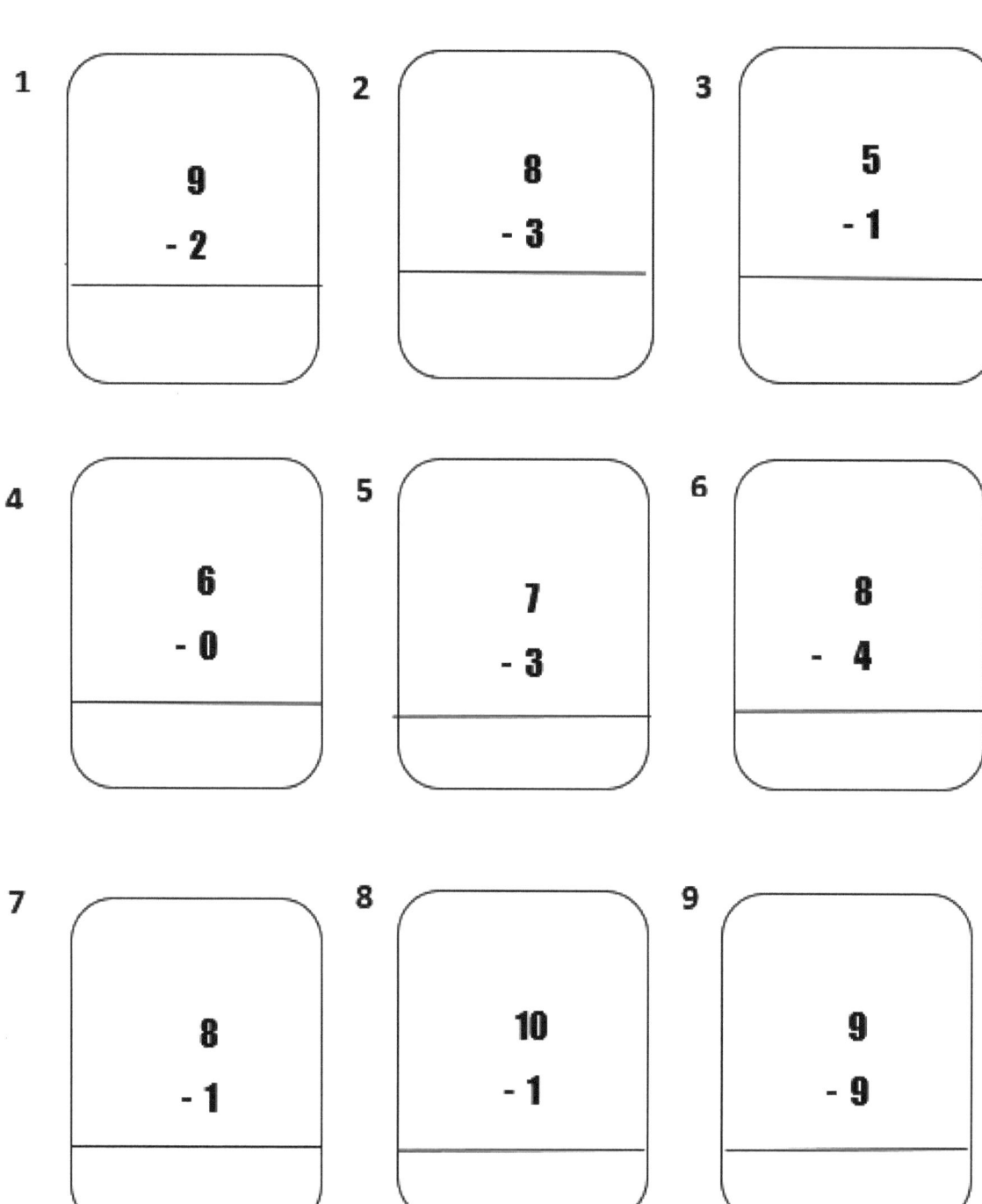

1
$$9$$
$$- 2$$

2
$$8$$
$$- 3$$

3
$$5$$
$$- 1$$

4
$$6$$
$$- 0$$

5
$$7$$
$$- 3$$

6
$$8$$
$$- 4$$

7
$$8$$
$$- 1$$

8
$$10$$
$$- 1$$

9
$$9$$
$$- 9$$

LET US SUBTRACT THOSE NUMBERS

Name- _____ Class- _____ Date- _____

1

$$\begin{array}{r} 9 \\ -\ 7 \\ \hline \end{array}$$

2

$$\begin{array}{r} 8 \\ -\ 5 \\ \hline \end{array}$$

3

$$\begin{array}{r} 5 \\ -\ 4 \\ \hline \end{array}$$

4

$$\begin{array}{r} 6 \\ -\ 5 \\ \hline \end{array}$$

5

$$\begin{array}{r} 7 \\ -\ 6 \\ \hline \end{array}$$

6

$$\begin{array}{r} 8 \\ -\ 7 \\ \hline \end{array}$$

7

$$\begin{array}{r} 8 \\ -\ 4 \\ \hline \end{array}$$

8

$$\begin{array}{r} 10 \\ -\ 3 \\ \hline \end{array}$$

9

$$\begin{array}{r} 9 \\ -\ 7 \\ \hline \end{array}$$

LET US SUBTRACT THOSE NUMBERS

Name- _____ Class- _____ Date- _____

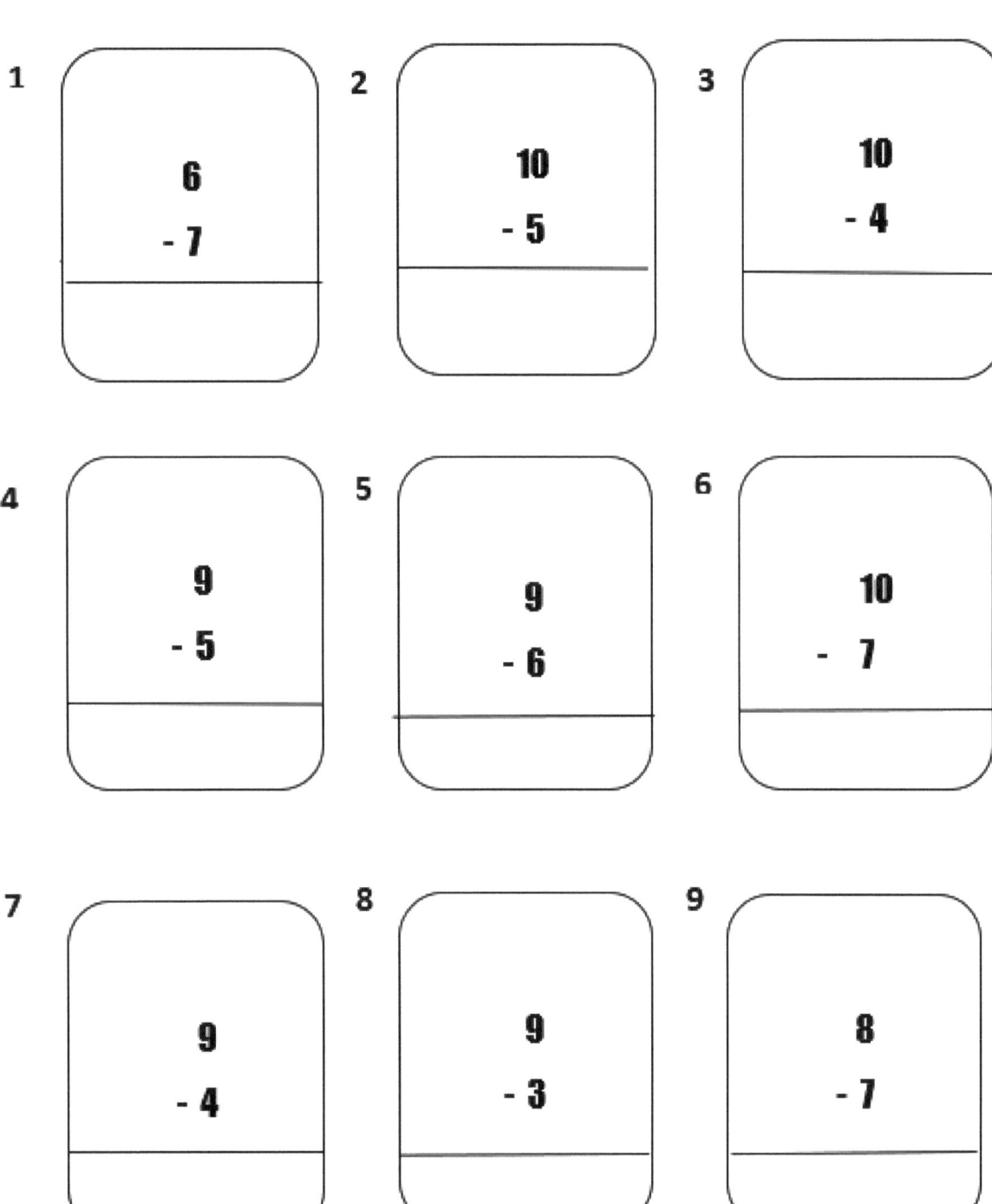

1.
$$\begin{array}{r} 6 \\ -7 \\ \hline \end{array}$$

2.
$$\begin{array}{r} 10 \\ -5 \\ \hline \end{array}$$

3.
$$\begin{array}{r} 10 \\ -4 \\ \hline \end{array}$$

4.
$$\begin{array}{r} 9 \\ -5 \\ \hline \end{array}$$

5.
$$\begin{array}{r} 9 \\ -6 \\ \hline \end{array}$$

6.
$$\begin{array}{r} 10 \\ -7 \\ \hline \end{array}$$

7.
$$\begin{array}{r} 9 \\ -4 \\ \hline \end{array}$$

8.
$$\begin{array}{r} 9 \\ -3 \\ \hline \end{array}$$

9.
$$\begin{array}{r} 8 \\ -7 \\ \hline \end{array}$$

LET US SUBTRACT THOSE NUMBERS

Name- _____ Class- _____ Date- _____

1
$$\begin{array}{r} 6 \\ -3 \\ \hline \end{array}$$

2
$$\begin{array}{r} 8 \\ -6 \\ \hline \end{array}$$

3
$$\begin{array}{r} 9 \\ -1 \\ \hline \end{array}$$

4
$$\begin{array}{r} 5 \\ -4 \\ \hline \end{array}$$

5
$$\begin{array}{r} 7 \\ -6 \\ \hline \end{array}$$

6
$$\begin{array}{r} 8 \\ -3 \\ \hline \end{array}$$

7
$$\begin{array}{r} 9 \\ -5 \\ \hline \end{array}$$

8
$$\begin{array}{r} 6 \\ -2 \\ \hline \end{array}$$

9
$$\begin{array}{r} 7 \\ -4 \\ \hline \end{array}$$

LET US SUBTRACT THOSE NUMBERS

Name- _____ Class- _____ Date- _____

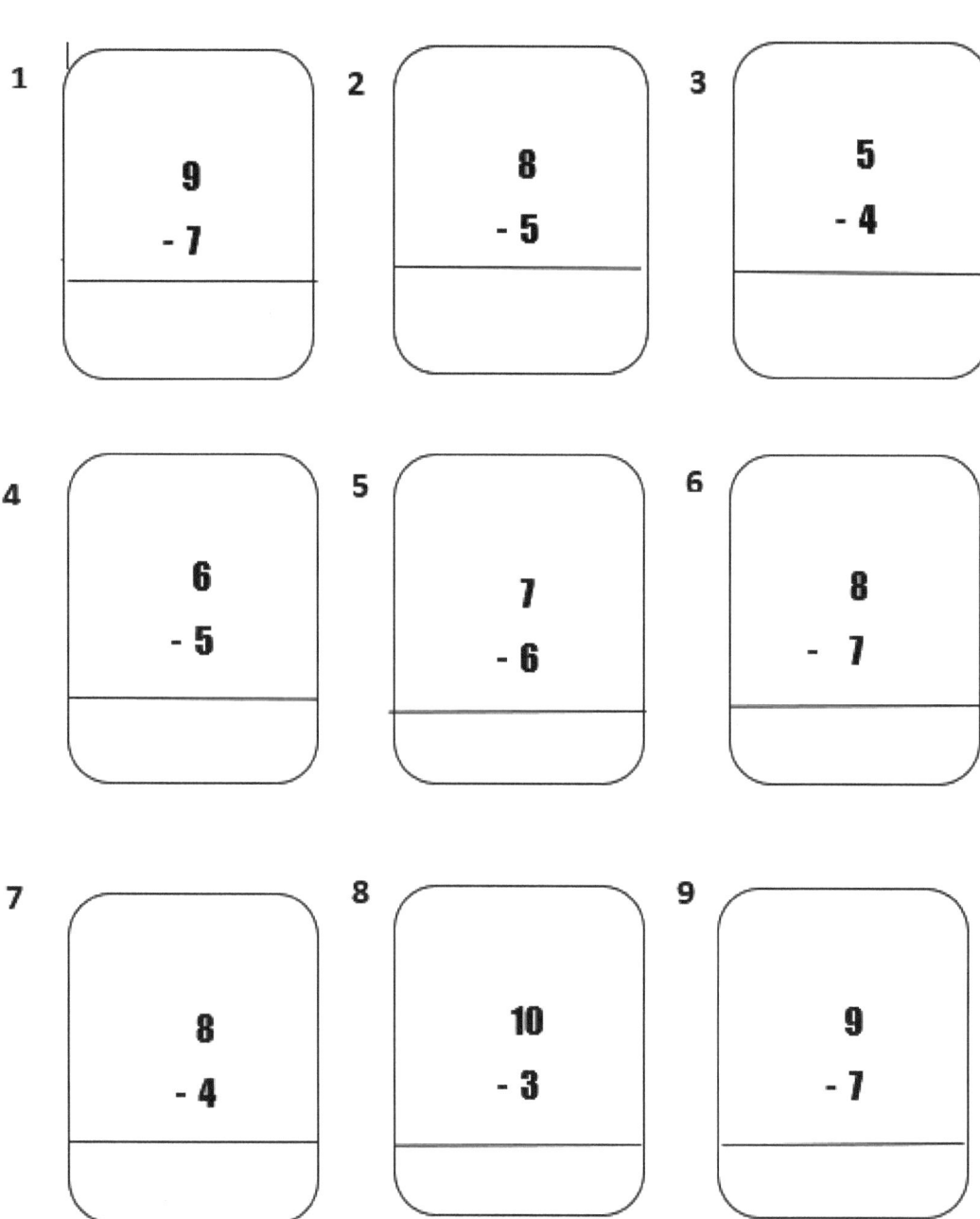

1.
```
   9
 - 7
```

2.
```
   8
 - 5
```

3.
```
   5
 - 4
```

4.
```
   6
 - 5
```

5.
```
   7
 - 6
```

6.
```
   8
 - 7
```

7.
```
   8
 - 4
```

8.
```
   10
 - 3
```

9.
```
   9
 - 7
```

- 2 X 4 = _ _ _ _ _
- 5 X 3 = _ _ _ _ _
- 1 X 7 = _ _ _ _ _
- 6 X 2 = _ _ _ _ _
- 3 X 6 = _ _ _ _ _
- 9 X 1 = _ _ _ _ _

- 4 X 3 = _ _ _ _ _
- 7 X 2 = _ _ _ _ _
- 8 X 1 = _ _ _ _ _
- 2 X 6 = _ _ _ _ _
- 5 X 4 = _ _ _ _ _
- 1 X 9 = _ _ _ _ _
- 7 X 1 = _ _ _ _ _
- 3 X 5 = _ _ _ _ _

- 6 X 3 = _ _ _ _ _
- 4 X 1 = _ _ _ _ _
- 2 X 7 = _ _ _ _ _
- 9 X 3 = _ _ _ _ _
- 8 X 5 = _ _ _ _ _
- 1 X 8 = _ _ _ _ _
- 6 X 4 = _ _ _ _ _
- 3 X 2 = _ _ _ _ _

- 5 X 7 = _ _ _ _ _
- 9 X 2 = _ _ _ _ _
- 1 X 6 = _ _ _ _ _
- 8 X 3 = _ _ _ _ _
- 4 X 6 = _ _ _ _ _
- 2 X 9 = _ _ _ _ _
- 7 X 3 = _ _ _ _ _
- 5 X 1 = _ _ _ _ _

- 3 X 7 = _ _ _ _ _
- 6 X 1 = _ _ _ _ _
- 9 X 4 = _ _ _ _ _
- 2 X 3 = _ _ _ _ _
- 1 X 4 = _ _ _ _ _
- 8 X 7 = _ _ _ _ _
- 7 X 6 = _ _ _ _ _

- 4 X 5 = _ _ _ _
- 3 X 4 = _ _ _ _
- 1 X 2 = _ _ _ _
- 9 X 6 = _ _ _ _
- 6 X 7 = _ _ _ _
- 2 X 5 = _ _ _ _

- 8 X 4 = ____
- 4 X 8 = ____
- 3 X 9 = ____
- 7 X 8 = ____
- 1 X 3 = ____
- 5 X 6 = ____
- 2 X 5 = ____

- 3 X 4 = _ _ _ _
- 7 X 2 = _ _ _ _
- 4 X 7 = _ _ _ _
- 1 X 9 = _ _ _ _
- 6 X 3 = _ _ _ _
- 9 X 1 = _ _ _ _
- 5 X 2 = _ _ _ _
- 3 X 7 = _ _ _ _

- 6 X 1 = _ _ _ _
- 2 X 6 = _ _ _ _
- 9 X 3 = _ _ _ _
- 7 X 4 = _ _ _ _
- 1 X 8 = _ _ _ _
- 3 X 3 = _ _ _ _
- 4 X 2 = _ _ _ _
- 5 X 6 = _ _ _ _

- 1 X 4 = ____
- 8 X 1 = ____
- 2 X 7 = ____
- 6 X 5 = ____
- 9 X 2 = ____
- 4 X 6 = ____
- 3 X 5 = ____
- 1 X 3 = ____

- 3 X 4 = ____
- 7 X 2 = ____
- 4 X 7 = ____
- 1 X 9 = ____
- 6 X 3 = ____
- 9 X 1 = ____
- 5 X 2 = ____
- 3 X 7 = ____

- $8 \div 2 = \_\_\_\_\_$
- $6 \div 3 = \_\_\_\_\_$
- $9 \div 3 = \_\_\_\_\_$
- $10 \div 2 = \_\_\_\_\_$
- $7 \div 1 = \_\_\_\_\_$
- $4 \div 2 = \_\_\_\_\_$
- $5 \div 1 = \_\_\_\_\_$

- 3 ÷ 1 = _ _ _ _
- 2 ÷ 2 = _ _ _ _
- 1 ÷ 1 = _ _ _ _
- 9 ÷ 9 = _ _ _ _
- 6 ÷ 2 = _ _ _ _
- 8 ÷ 4 = _ _ _ _

- $4 \div 1 =$ _ _ _ _
- $3 \div 2 =$ _ _ _ _
- $7 \div 7 =$ _ _ _ _
- $1 \div 1 =$ _ _ _ _
- $5 \div 5 =$ _ _ _ _
- $8 \div 1 =$ _ _ _ _
- $6 \div 6 =$ _ _ _ _

- $9 \div 1 = $ _ _ _ _
- $2 \div 1 = $ _ _ _ _
- $7 \div 2 = $ _ _ _ _
- $3 \div 3 = $ _ _ _ _
- $1 \div 1 = $ _ _ _ _
- $8 \div 8 = $ _ _ _ _

- $6 \div 1 =$ _ _ _ _ _
- $4 \div 1 =$ _ _ _ _ _
- $3 \div 1 =$ _ _ _ _ _
- $2 \div 2 =$ _ _ _ _ _
- $5 \div 1 =$ _ _ _ _ _
- $6 \div 2 =$ _ _ _ _ _
- $9 \div 1 =$ _ _ _ _ _

- 6 ÷ 1 = _ _ _ _ _
- 4 ÷ 1 = _ _ _ _ _
- 3 ÷ 1 = _ _ _ _ _
- 2 ÷ 2 = _ _ _ _ _
- 5 ÷ 1 = _ _ _ _ _
- 6 ÷ 2 = _ _ _ _ _
- 9 ÷ 1 = _ _ _ _ _

- $6 \div 1 = $ _____
- $10 \div 2 = $ _____
- $18 \div 3 = $ _____
- $12 \div 4 = $ _____
- $15 \div 5 = $ _____
- $8 \div 2 = $ _____
- $20 \div 5 = $ _____

- $6 \div 1 =$ _ _ _ _ _
- $10 \div 2 =$ _ _ _ _
- $18 \div 3 =$ _ _ _ _
- $12 \div 4 =$ _ _ _ _ _
- $15 \div 5 =$ _ _ _ _ _
- $8 \div 2 =$ _ _ _ _ _
- $20 \div 5 =$ _ _ _ _

- 16 ÷ 4 = _____
- 21 ÷ 7 = _____
- 24 ÷ 6 = _____
- 28 ÷ 7 = _____
- 14 ÷ 2 = _____
- 36 ÷ 9 = _____
- 30 ÷ 5 = _____

- 16 ÷ 4 = _ _ _ _
- 32 ÷ 8 = _ _ _ _
- 25 ÷ 5 = _ _ _ _
- 9 ÷ 3 = _ _ _ _
- 6 ÷ 3 = _ _ _ _
- 16 ÷ 4 = _ _ _
- 12 ÷ 2 = _ _ _

- $12 \div 2 =$ _ _ _ _
- $10 \div 2 =$ _ _ _ _
- $12 \div 3 =$ _ _ _ _
- $18 \div 3 =$ _ _ _ _
- $8 \div 2 =$ _ _ _ _
- $4 \div 2 =$ _ _ _ _

ABOUT THE AUTHOR

Nisha is an educational professional with a fervor for storytelling and a background in science. She loves storytelling in her classrooms and loves to work on the social-emotional learning of young minds. She loves to create helpful content for learning and reading. She believes in making this world a better place by sensitizing people with better teaching-learning-knowing processes.